The Rise of Gliese 12b

The mystery behind the discovery of a new planet and what it means for the solar system

By

Barbs Walters

Copyright © 2024, by Barbs Walters.

All rights reserved. No part of this book may be reproduced or transmitted in any form or by any means, electronic or mechanical, including photocopying, recording, or any information storage and retrieval system, without permission in writing from the copyright owner, except for brief quotations in critical reviews and articles.

Table of contents

The Rise of Gliese 12b
What to know of the newly discovered planet
By
Des Waters
Copyright © 2024, by Des Waters.
Table of contents
Chapter 1: Introduction to Exoplanet Discovery
    1.1 The Search for Habitable Worlds
    1.2 Importance of Exoplanet Research
Chapter 2: Gliese 12b: Discovery and Significance
    2.1 The Discovery of Gliese 12b
    2.2 Significance of Gliese 12b
Chapter 3: The Star System: Gliese 12
    3.1 Characteristics of Red Dwarf Stars
    3.2 The Constellation Pisces and Gliese 12
Chapter 4: Physical and Orbital Characteristics of Gliese 12b
    4.1 Physical Properties of Gliese 12b
    4.2 Orbital Dynamics
Chapter 5: Habitability Potential
    5.1 The Habitable Zone
    5.2 Comparative Analysis with Earth and Venus
Chapter 6: Technological and Methodological Advances

6.1 Observational Techniques and Data Collection
6.2 Future Research and Technological Tools
Chapter 7: Implications and Future Exploration
7.1 Astrobiological and Scientific Implications
7.2 The Future of Exoplanet Exploration

Chapter 1: Introduction to Exoplanet Discovery

1.1 The Search for Habitable Worlds

The quest to discover planets beyond our solar system, known as exoplanets, has become one of the most exciting and rapidly evolving fields in astronomy. This journey began in earnest in the early 1990s, fundamentally altering our understanding of the universe and our place within it. The search for habitable worlds revolves around the compelling question: Are we alone in the universe? By finding and studying exoplanets, especially those that might harbour life, scientists hope to uncover answers that could redefine our perspective on life and its potential ubiquity.

Historically, the notion of planets orbiting stars other than our Sun was purely speculative. Philosophers and early astronomers pondered the existence of other worlds, but without empirical evidence, these ideas remained in the realm of conjecture. This changed dramatically with the advent of modern telescopic technology and observational techniques.

In 1992, the first confirmed detection of exoplanets came with the discovery of several terrestrial-mass planets orbiting the pulsar PSR B1257+12. These findings were groundbreaking but also perplexing, as pulsars are the remnants of supernovae, and their intense radiation seemed an unlikely cradle for life. However, this discovery set the stage for subsequent research, demonstrating that planets could exist in a wide variety of environments.

The next pivotal moment came in 1995 when Michel Mayor and Didier Queloz announced the discovery of 51 Pegasi b, a gas giant orbiting a Sun-like star. This milestone proved that planetary systems could form around stars similar to our own, igniting a surge of interest and research. Using the radial velocity method, which detects wobbles in a star's motion caused by gravitational tugs from orbiting planets, astronomers began uncovering a plethora of exoplanets, many of which defied previous expectations.

1.2 Importance of Exoplanet Research

The discovery of exoplanets is not merely a cataloguing exercise but a profound scientific endeavour that impacts various fields of study, from astrobiology to planetary science and cosmology. Understanding the diversity of planetary systems helps scientists piece together the processes that lead to planet formation, including our own solar system's history.

One of the most significant aspects of exoplanet research is its implications for the search for life. The conditions required for life as we know it—liquid water, a stable climate, and the right chemical ingredients—depend heavily on a planet's location relative to its star, known as the habitable zone. By identifying planets within this zone, researchers can prioritise targets for further study, increasing the chances of detecting signs of life.

The methods used to discover and study exoplanets have also driven technological advancements in telescopes and data analysis. Instruments like the Kepler Space Telescope and the Transiting Exoplanet Survey Satellite (TESS) have revolutionised our ability to detect and monitor distant worlds. Kepler, launched in 2009, used the transit method to observe minute dips in starlight caused by planets passing in front of their stars. This technique enabled the discovery of thousands of exoplanets, many of which are Earth-sized and located within their stars' habitable zones.

TESS, launched in 2018, continues this legacy by surveying the entire sky, focusing on stars closer to Earth. This mission aims to find exoplanets that are prime candidates for detailed study by future missions, such as the James Webb Space Telescope (JWST). JWST, with its powerful infrared capabilities, will analyse the atmospheres of exoplanets, searching for signs of water vapour, oxygen, methane, and other potential indicators of life.

The field of exoplanet research also intersects with planetary science, as studying these distant worlds provides context for understanding the planets within our own solar system. By comparing exoplanets with Earth, Mars, Venus, and other solar system bodies, scientists can explore how different environmental conditions affect planetary development and potential habitability.

Furthermore, the discovery of exoplanets has profound philosophical and cultural implications. The realisation that planets are common throughout the galaxy challenges our long-held views about the uniqueness of Earth. It raises questions about the potential for diverse forms of life and civilizations beyond our planet, prompting reflection on humanity's place in the cosmos.

This chapter provides an in-depth introduction to the discovery and study of exoplanets. It traces the historical development of the field, highlighting key discoveries and technological advancements that have transformed our understanding of planetary systems. By focusing on the methods used to detect and characterise exoplanets, the chapter underscores the importance of this research in the broader context of astronomy and astrobiology.

The subsequent sections will delve into the specifics of the recently discovered exoplanet Gliese 12b, examining its characteristics, the star it orbits, and its potential habitability. This detailed case study will illustrate the scientific processes and collaborative efforts involved in exoplanet research, offering insights into the cutting-edge techniques and future directions of this dynamic field.

The discovery of Gliese 12b is a testament to the progress made in exoplanet research. Located 40 light-years away in the constellation Pisces. Despite its proximity to its star, the planet falls within the habitable zone, where conditions might allow for liquid water—a crucial ingredient for life as we know it.

By studying planets like Gliese 12b, scientists aim to understand the factors that make a planet habitable and how these factors vary across different planetary systems. This knowledge not only enhances our understanding of the universe but also informs the search for life beyond Earth. The discovery of potentially habitable exoplanets fuels the hope that we might one day find evidence of life elsewhere in the cosmos, answering one of humanity's oldest and most profound questions.

Chapter 2: Gliese 12b: Discovery and Significance

2.1 The Discovery of Gliese 12b

The discovery of Gliese 12b marks a significant milestone in the ongoing quest to identify potentially habitable planets beyond our solar system. Located approximately 40 light-years away in the constellation Pisces, Gliese 12b orbits a red dwarf star named Gliese 12. Red dwarfs are the most common type of star in the Milky Way, characterised by their small size, low luminosity, and cooler temperatures compared to stars like our Sun. Despite these differences, red dwarfs have become focal points in exoplanetary studies due to their abundance and the relative ease of detecting planets around them.

Gliese 12b is of particular interest because it resides within the habitable zone of its host star—the region where conditions might be right for liquid water to exist on a planet's surface. This planet is slightly smaller than Earth but larger than Venus, suggesting that it might share some physical characteristics with terrestrial planets in our own solar system. The possibility that Gliese 12b could have Earth-like conditions makes it an exciting target for further study.

The discovery was made using data from NASA's Transiting Exoplanet Survey Satellite (TESS), which was launched in April 2018. TESS is designed to survey the entire sky, focusing on stars that are relatively close to Earth. By monitoring the brightness of thousands of stars, TESS identifies exoplanets through the transit method, which detects the slight dimming of a star as a planet passes in front of it. This method has proven highly effective in identifying exoplanets, particularly those orbiting smaller, dimmer stars like red dwarfs.

The Role of NASA's Transiting Exoplanet Survey Satellite (TESS)

TESS's mission is to find exoplanets around the brightest stars near Earth, providing prime targets for detailed study by future telescopes. The satellite is equipped with four wide-field cameras that allow it to observe a large portion of the sky, capturing changes in stellar brightness with remarkable precision. Each camera has a field of view that covers an area 24 degrees across, enabling TESS to survey large swathes of the sky simultaneously.

The discovery process for Gliese 12b began with the identification of periodic dips in the brightness of Gliese 12, indicative of a planetary transit. These observations were then subjected to rigorous analysis to rule out other potential causes, such as stellar activity or binary star interactions. Once confirmed as a planet, Gliese 12b was catalogued, and its orbital and physical characteristics were studied in greater detail.

TESS's ability to detect small, Earth-sized planets in the habitable zones of their stars represents a significant advancement over previous missions. Its broad survey capabilities and focus on nearby stars mean that any planets it discovers are excellent candidates for follow-up observations with more powerful instruments, such as the James Webb Space Telescope (JWST). By identifying and characterising planets like Gliese 12b, TESS helps to build a comprehensive catalogue of potentially habitable worlds, setting the stage for future discoveries that could reveal much about the conditions necessary for life.

2.2 Significance of Gliese 12b

Gliese 12b stands out among the myriad of exoplanets discovered to date due to its size, location, and potential habitability. Most exoplanets found thus far fall into two categories: gas giants like Jupiter and Saturn, or much smaller, rocky worlds. Finding a planet that is intermediate in size between Earth and Venus is relatively rare, making Gliese 12b a valuable addition to the growing list of known exoplanets.

When comparing Gliese 12b to other exoplanets, it is essential to consider its host star. Red dwarfs like Gliese 12 are cooler and less luminous than Sun-like stars, which means their habitable zones are much closer to the star. This proximity can have significant implications for the planet's climate and atmospheric conditions. For example, planets in close orbits around red dwarfs are often tidally locked, meaning one side of the planet perpetually faces the star while the other remains in darkness. This could result in extreme temperature differences between the two hemispheres.

In contrast, many of the exoplanets discovered around Sun-like stars are located further from their host stars, with orbital periods measured in months or years rather than days. These planets often have more Earth-like day-night cycles, which could lead to more moderate and stable climates. However, the intense stellar activity common in red dwarfs, such as frequent flares and high levels of radiation, could pose challenges for habitability that planets around Sun-like stars might not face.

Despite these differences, the discovery of Gliese 12b adds valuable data to our understanding of the diversity of planetary systems. By studying planets around various types of stars, astronomers can develop more comprehensive models of planet formation and evolution. This, in turn, helps refine the criteria used to identify potentially habitable worlds and enhances our understanding of the factors that contribute to planetary habitability.

Implications for the Study of Habitable Zones

The concept of the habitable zone is central to the search for life beyond Earth. Defined as the region around a star where conditions might be right for liquid water to exist, the habitable zone varies depending on the star's size and luminosity. For Gliese 12, a red dwarf, the habitable zone is much closer to the star than it would be for a Sun-like star. Gliese 12b's location within this zone makes it a prime candidate for studying the conditions that could support life.

One of the key questions scientists aim to answer is whether Gliese 12b has an atmosphere and, if so, what its composition might be. The presence of an atmosphere is crucial for regulating a planet's temperature and protecting it from harmful radiation. On Earth, the atmosphere plays a vital role in maintaining the planet's climate and enabling the water cycle. If Gliese 12b has a similar atmosphere, it could potentially support liquid water and, by extension, life.

However, detecting and analysing the atmosphere of a distant exoplanet is a challenging task. Future missions, particularly the James Webb Space Telescope, are expected to play a crucial role in this area. JWST's advanced spectroscopic capabilities will allow scientists to analyse the light passing through Gliese 12b's atmosphere during transits, identifying the chemical signatures of various gases. This information can provide insights into the planet's climate, potential for water, and overall habitability.

The study of habitable zones also involves understanding the star's activity. Red dwarfs are known for their high levels of stellar activity, including frequent flares that can emit intense bursts of radiation. These flares have the potential to strip away a planet's atmosphere, making it less hospitable for life. Observing and modelling these interactions are essential for assessing the true habitability of planets like Gliese 12b.

Moreover, the discovery of Gliese 12b contributes to the broader effort to understand the conditions that allow a planet to remain habitable over geological timescales. By comparing Gliese 12b with planets in our solar system, such as Earth and Venus, scientists can explore why some planets maintain stable climates while others experience runaway greenhouse effects or other climatic extremes. These comparisons can shed light on the processes that influence planetary habitability and the potential for life on other worlds.

Chapter 3: The Star System: Gliese 12

3.1 Characteristics of Red Dwarf Stars

Red dwarf stars, also known as M-dwarfs, are the smallest and most numerous type of star in the Milky Way galaxy. These stars have masses ranging from about 0.08 to 0.6 times that of our Sun and radii between 0.1 and 0.7 solar radii. Their low mass and size result in lower temperatures and luminosities. Typically, red dwarfs have surface temperatures between 2,500 and

4,000 Kelvin, which is much cooler compared to the Sun's surface temperature of approximately 5,778 Kelvin. Due to these lower temperatures, red dwarfs emit light primarily in the red and infrared parts of the electromagnetic spectrum, giving them their distinctive hue.

The internal structure of red dwarfs also differs significantly from that of more massive stars. Red dwarfs are fully convective, meaning that energy produced in their cores is transported to the surface by convection currents rather than radiation. This convective process ensures that hydrogen fuel is efficiently mixed throughout the star, allowing red dwarfs to burn their fuel much more slowly than larger stars. As a result, red dwarfs have incredibly long lifespans, often exceeding tens of billions of years, and in some cases, they can remain in the main sequence phase for trillions of years.

Red dwarfs are also known for their magnetic activity. They possess strong magnetic fields generated by convective motions within their interiors. This magnetic activity can lead to frequent and intense stellar flares—sudden releases of energy that can dramatically increase the star's brightness for short periods. These flares can emit harmful ultraviolet and X-ray radiation, which poses challenges for the habitability of planets orbiting red dwarfs. The magnetic activity can also drive stellar winds, streams of charged particles that can strip away the atmospheres of nearby planets, further complicating the prospects for habitability.

Despite these challenges, the unique characteristics of red dwarfs make them intriguing subjects for exoplanet studies. Their small size and lower luminosity mean that planets in their habitable zones—where liquid water could potentially exist—are much closer to the star. This proximity enhances the transit method's effectiveness, where astronomers detect planets by observing periodic dips in a star's brightness caused by planets passing in front of it. The greater relative dimming effect makes it easier to detect Earth-sized planets around red dwarfs compared to larger, brighter stars.

Importance of Red Dwarfs in Exoplanet Studies

The abundance of red dwarf stars in the galaxy makes them a focal point for exoplanet research. Red dwarfs constitute about 70-80% of all stars in the Milky Way, suggesting that a significant number of exoplanets may orbit these stars. This prevalence increases the likelihood of finding potentially habitable planets within their systems, making red dwarfs prime targets in the search for extraterrestrial life.

The discovery of potentially habitable planets around red dwarfs, such as Gliese 12b, has significant implications for our understanding of planetary formation and the conditions necessary for life. Since red dwarfs are so common, understanding the habitability of their planetary systems could provide insights into the broader potential for life throughout the galaxy. If planets around red dwarfs can support life, it would imply that habitable environments are much more widespread than previously thought.

Furthermore, the study of planets around red dwarfs helps astronomers test theories of planetary formation and migration. The close-in habitable zones of red dwarfs provide unique laboratories for studying how planets form and evolve in environments with intense stellar radiation and magnetic activity. Observations of these systems can reveal how planets develop atmospheres, regulate their climates, and potentially support life despite the challenges posed by their host stars.

Red dwarfs' long lifespans also offer stability that might be conducive to the development of life. Unlike larger stars that can burn out or explode in supernovae relatively quickly on cosmic timescales, red dwarfs provide a stable energy source for billions or even trillions of years. This longevity offers ample time for life to potentially arise and evolve on orbiting planets, given the right conditions.

In addition to these scientific advantages, the proximity of many red dwarf systems to Earth makes them accessible targets for detailed study. Many of the nearest stars to our solar system are red dwarfs, including

Proxima Centauri, the closest known star to the Sun. These nearby red dwarfs provide opportunities for high-resolution observations with current and future telescopes, allowing scientists to study their planets in greater detail and search for signs of habitability and life.

3.2 The Constellation Pisces and Gliese 12

The constellation Pisces is one of the 88 modern constellations recognized by the International Astronomical Union (IAU). It is located in the northern sky and represents two fish tied together by their tails. Pisces is part of the zodiac, a group of constellations through which the Sun, Moon, and planets appear to move over the course of the year. The zodiac has historical and astrological significance, with Pisces traditionally associated with water and the spring equinox.

Pisces is a relatively faint constellation, lacking any particularly bright stars. However, it is rich in deep-sky objects and has a number of notable features. One of the most famous objects within Pisces is the spiral galaxy M74, also known as the Phantom Galaxy. This galaxy is about 32 million light-years away and is known for its well-defined spiral arms and bright nucleus, making it a popular target for amateur astronomers.

The constellation also contains a number of double stars and variable stars, providing diverse targets for both professional and amateur observations. Despite its relatively low prominence in the night sky, Pisces holds significance due to its position along the ecliptic and its role in the zodiac.

Position and Characteristics of Gliese 12 within Pisces

Gliese 12 is a red dwarf star located within the boundaries of the constellation Pisces. Like many other red dwarfs, it is not visible to the naked eye due to its low luminosity. However, it has become an object of interest due to the discovery of its orbiting exoplanet, Gliese 12b. Understanding the position and characteristics of Gliese 12 within Pisces provides context for its study and highlights its importance in the broader field of exoplanet research.

Gliese 12 lies at a distance of approximately 40 light-years from Earth. This relatively close proximity in astronomical terms makes it an accessible target for observation and study. The star's position within Pisces situates it within a region of the sky that is well-studied and charted, allowing astronomers to accurately pinpoint its location and monitor its behaviour over time.

The characteristics of Gliese 12, typical of red dwarfs, include its low mass, small size, and cool temperature. These properties influence the conditions on its orbiting planet, Gliese 12b, and play a crucial role in determining the planet's potential habitability. The star's low luminosity means that its habitable zone is much closer than that of a star like the Sun, resulting in shorter orbital periods for planets within this zone. In the case of Gliese 12b, the planet completes an orbit every 12.8 days, positioning it well within the habitable zone.

Understanding the specific characteristics of Gliese 12, such as its magnetic activity and flare frequency, is essential for assessing the habitability of Gliese 12b. Red dwarfs are known for their high levels of stellar activity, which can have significant effects on orbiting planets. Frequent flares can strip away planetary atmospheres, exposing the surface to harmful radiation and reducing the potential for life. Monitoring Gliese 12's activity helps scientists predict the environmental conditions on Gliese 12b and evaluate its suitability for supporting life.

The study of Gliese 12 within the context of its position in Pisces also contributes to our broader understanding of red dwarf systems. By comparing Gliese 12 to other red dwarfs in different regions of the sky, astronomers can identify patterns and variations in stellar behaviour and planetary formation. This comparative approach enhances our knowledge of how different types of stars influence the development of their planetary systems and the potential for habitability.

Chapter 4: Physical and Orbital Characteristics of Gliese 12b

4.1 Physical Properties of Gliese 12b

Gliese 12b is an intriguing exoplanet due to its size, which places it between Earth and Venus in terms of planetary dimensions. With a radius approximately 1.1 times that of Earth, Gliese 12b is slightly larger, suggesting a planet that could have similarities to our terrestrial worlds. Its mass, estimated to be around 1.5 times that of Earth, further indicates that it is a rocky planet rather than a gas giant. The combination of its size and mass suggests that Gliese 12b has a substantial gravitational field, which could allow it to retain a significant atmosphere.

The planet's composition is inferred from its density, which has been estimated using its radius and mass. Given these parameters, Gliese 12b likely has a rocky composition similar to Earth, with a core composed of iron and nickel, surrounded by a mantle of silicate minerals. This rocky nature is crucial for habitability since it suggests the presence of a solid surface where liquid water, a key ingredient for life, could potentially exist.

Additionally, the planet's relatively high density implies that it does not possess the thick gaseous envelope characteristic of mini-Neptunes or other small gas giants. This rocky composition places Gliese 12b firmly in the category of terrestrial planets, making it a prime candidate for studying the conditions that might support life.

Surface Temperature and Atmospheric Conditions

The surface temperature of Gliese 12b, assuming it lacks a significant atmosphere, is estimated to be around 107 degrees Fahrenheit (42 degrees Celsius). This temperature is based on its distance from its host star, Gliese 12, and the star's relatively low luminosity. While this temperature is higher than the average on Earth, it is still within the range where liquid water could exist, especially if the planet has mechanisms to moderate its climate.

The presence and nature of an atmosphere on Gliese 12b are still unknown, but they are crucial for determining the planet's habitability. An atmosphere can regulate a planet's temperature, distribute heat around the planet, and protect the surface from harmful stellar radiation. If Gliese 12b possesses a thick atmosphere with greenhouse gases like carbon dioxide, water vapour, or methane, it could have a surface temperature much higher than the estimated 107 degrees Fahrenheit due to a runaway greenhouse effect, similar to what we observe on Venus.

Conversely, if Gliese 12b has a thin or no atmosphere, its surface conditions could be harsh, with extreme temperature fluctuations between day and night. The lack of an atmosphere would also expose the surface to higher levels of stellar radiation, particularly given the frequent flaring activity of red dwarf stars like Gliese 12.

Current observational techniques, such as those that will be employed by the James Webb Space Telescope (JWST), aim to determine the atmospheric composition of exoplanets like Gliese 12b. By analysing the starlight that filters through the planet's atmosphere during transits, scientists can identify the chemical signatures of various gases. This spectroscopic analysis could reveal whether Gliese 12b has an atmosphere rich in water vapour or other greenhouse gases, or if it is more akin to the airless surface of Mercury.

Understanding the atmospheric conditions of Gliese 12b is essential not only for assessing its current habitability but also for drawing comparisons with Earth and Venus. These comparisons can help elucidate why Earth remains habitable while Venus has become a hostile environment, offering insights into the factors that influence planetary climates and the potential for life on other worlds.

4.2 Orbital Dynamics

Gliese 12b orbits its host star, Gliese 12, at a remarkably close distance, completing an orbit every 12.8 days. This short orbital period places Gliese 12b well within the habitable zone of its red dwarf star, a region where conditions might allow for the presence of liquid water on the planet's surface. The proximity of the habitable zone to the star in red dwarf systems is a result of the star's lower luminosity compared to stars like our Sun. Despite its close orbit, the reduced luminosity of Gliese 12 means that the planet receives a level of stellar radiation that could permit moderate surface temperatures, assuming an atmosphere is present to regulate the climate.

The tight orbit of Gliese 12b has significant implications for the planet's rotational dynamics. It is highly likely that Gliese 12b is tidally locked, meaning one side of the planet always faces the star while the other remains in perpetual darkness. Tidal locking is a common outcome for planets in close orbits around their stars due to gravitational interactions over long timescales. This configuration leads to extreme temperature contrasts between the day side and the night side of the planet, potentially affecting atmospheric circulation patterns and climate stability.

The day side of a tidally locked planet could experience intense heating, while the night side could be exceedingly cold. However, if Gliese 12b has a thick atmosphere, heat redistribution via atmospheric winds could moderate these temperature extremes, creating a more stable and habitable environment. Understanding how heat is transported around the planet is crucial for assessing the potential for life and the presence of liquid water.

Implications for Habitability and Climate

The close orbit of Gliese 12b within the habitable zone of its star suggests that it is in a region where liquid water could exist, a key criterion for habitability. However, several factors must be considered when evaluating the planet's potential to support life.

Firstly, the tidal locking of Gliese 12b means that the climate could be highly asymmetric. The perpetual daylight on one side could lead to significant heating, potentially causing surface temperatures to rise above the boiling point of water unless atmospheric mechanisms can effectively distribute heat. The perpetual night on the opposite side could result in extremely cold temperatures, potentially freezing any water present. If the planet has an atmosphere, especially one with a significant greenhouse effect, it could help balance these temperature extremes, creating a habitable twilight zone along the terminator—the line separating day and night.

Secondly, the magnetic activity of the host star, Gliese 12, plays a crucial role in the habitability of Gliese 12b. Red dwarf stars are known for their frequent and intense stellar flares, which can emit high levels of ultraviolet and X-ray radiation. These flares can erode planetary atmospheres and expose the surface to harmful radiation, posing challenges for the development and sustainability of life. The presence of a strong magnetic field on Gliese 12b could offer some protection by deflecting charged particles and mitigating atmospheric loss.

Thirdly, the atmospheric composition of Gliese 12b is a pivotal factor in determining its habitability. An atmosphere rich in greenhouse gases could create a runaway greenhouse effect, leading to Venus-like conditions with extremely high surface temperatures. On the other hand, an atmosphere with a balanced composition of gases could support a stable climate, potentially allowing liquid water to persist on the surface. The presence of water vapour in the atmosphere would be a strong indicator of potential habitability, as it suggests the presence of surface or subsurface water.

The study of Gliese 12b's orbit and its implications for climate is also essential for understanding the broader context of habitable environments around red dwarf stars. By comparing Gliese 12b with other known exoplanets in similar systems, scientists can identify patterns and anomalies that provide insights into planetary formation and evolution. This comparative analysis helps refine models of habitability and improves our ability to identify other potentially habitable planets.

In addition to observational studies, theoretical modelling plays a crucial role in understanding the climate and habitability of Gliese 12b. Climate models that simulate the planet's atmospheric dynamics, heat distribution, and potential weather patterns can provide valuable predictions about its environment. These models can be adjusted based on observational data from telescopes and space missions, creating a comprehensive picture of the planet's climate.

Future missions and technological advancements will continue to enhance our understanding of Gliese 12b and similar exoplanets. The James Webb Space Telescope, with its advanced spectroscopic capabilities, will be able to analyse the atmospheric composition of exoplanets in unprecedented detail, potentially identifying biomarkers and other signs of habitability. Additionally, missions focused on direct imaging of exoplanets will allow scientists to observe these distant worlds more directly, providing further insights into their physical and atmospheric properties.

Chapter 5: Habitability Potential

5.1 The Habitable Zone

The concept of the habitable zone, often referred to as the "Goldilocks Zone," is a critical factor in the search for life beyond Earth. This zone represents the range of distances from a star within which planetary conditions might allow for liquid water to exist on a planet's surface. Water is a fundamental ingredient for life as we know it, making the habitable zone a key target for exoplanetary research.

The habitable zone's distance from a star depends on several factors, including the star's luminosity and temperature. For a star like our Sun, the habitable zone typically lies between 0.95 and 1.37 astronomical units (AU) from the star. However, for cooler, less luminous stars like red dwarfs, the habitable zone is much closer in, often within a fraction of an AU.

The importance of the habitable zone lies in its potential to support liquid water, which is essential for the chemistry of life. Water acts as a solvent, facilitating the complex biochemical reactions necessary for life. Planets within this zone, especially those that are Earth-sized and rocky, are prime candidates for further study in the quest to find life elsewhere in the universe.

Moreover, the habitable zone is not a static region. It can change over time due to stellar evolution, planetary atmosphere dynamics, and other factors. Understanding these dynamics is crucial for assessing the long-term habitability of exoplanets. For instance, a planet might start within the habitable zone but move out of it due to changes in its star's luminosity or its own orbital dynamics.

Analysis of Gliese 12b's Position within Its Habitable Zone

Gliese 12b orbits a red dwarf star named Gliese 12, situated within the constellation

Pisces. The star is significantly smaller and cooler than our Sun, with approximately 27% of the Sun's size and 60% of its temperature. Due to these characteristics, the habitable zone around Gliese 12 is much closer to the star compared to the Sun's habitable zone.

Gliese 12b orbits its star every 12.8 days at a distance that places it within the habitable zone of Gliese 12. This proximity allows the planet to receive sufficient stellar radiation to potentially support liquid water on its surface, despite the lower overall luminosity of the star. The planet's position within this zone makes it an intriguing candidate for the search for life.

The exact location of Gliese 12b within the habitable zone affects its potential climate and habitability. Being closer to the inner edge of the habitable zone might subject the planet to higher temperatures, which could lead to a runaway greenhouse effect if the atmosphere is rich in greenhouse gases. Conversely, if it lies closer to the outer edge, the planet might be cooler, which could challenge the presence of liquid water unless there are significant greenhouse gases to retain heat.

The dynamic nature of red dwarf stars, which often exhibit significant magnetic activity and stellar flares, further complicates the assessment of Gliese 12b's habitability. Frequent flares could strip away the planet's atmosphere or subject its surface to intense radiation, posing challenges for the development and sustainability of life.

Understanding Gliese 12b's exact position and the factors influencing its climate requires detailed observational data and advanced modelling. Instruments like the James Webb Space Telescope (JWST) will play a crucial role in providing the data needed to analyse the planet's atmosphere and surface conditions, offering insights into its potential to support life.

5.2 Comparative Analysis with Earth and Venus

Comparing Gliese 12b with Earth provides a framework for assessing its habitability. Earth is the only known planet to support life, making it a benchmark for habitability studies. Several key factors contribute to Earth's ability to support life, including its stable climate, presence of liquid water, protective atmosphere, and magnetic field.

For Gliese 12b to have Earth-like conditions, it would need a stable climate that allows for the presence of liquid water over long periods. This requires a balanced atmospheric composition, with sufficient greenhouse gases to retain heat but not so much as to trigger a runaway greenhouse effect. Water vapour, carbon dioxide, and methane are critical components in this balance, contributing to the greenhouse effect that keeps Earth warm enough to support life.

The potential for liquid water on Gliese 12b depends on its surface temperature and atmospheric conditions. If the planet has an atmosphere that can moderate temperature extremes, liquid water could exist on its surface. The estimated surface temperature of 107 degrees Fahrenheit (42 degrees Celsius) suggests that, without an atmosphere, conditions might be too harsh for liquid water to persist. However, if Gliese 12b has a substantial atmosphere, it could stabilise temperatures and create pockets where water can remain liquid.

Another critical factor is the planet's magnetic field. Earth's magnetic field protects its atmosphere from solar wind and cosmic radiation, which could otherwise strip away the atmosphere and water. If Gliese 12b has a strong magnetic field, it could similarly protect its atmosphere, enhancing its habitability potential.

Understanding these factors requires detailed observational data, which can reveal the presence of water vapour and other gases in Gliese 12b's atmosphere. Spectroscopic analysis, particularly using the capabilities of the JWST, will be essential in this endeavour.

Lessons from Venus and the Greenhouse Effect

Venus provides a cautionary example of how a planet within the habitable zone can become inhospitable due to the runaway greenhouse effect. Despite being similar in size and composition to Earth, Venus has a thick atmosphere dominated by carbon dioxide, leading to surface temperatures around 900 degrees Fahrenheit (475 degrees Celsius), hot enough to melt lead.

The study of Venus illustrates the critical role of atmospheric composition in determining a planet's climate. The high concentration of carbon dioxide on Venus traps heat, creating a feedback loop that drives temperatures ever higher. This runaway greenhouse effect makes Venus an inhospitable world, despite its location within the habitable zone.

Comparing Gliese 12b to Venus helps highlight the delicate balance required for habitability. If Gliese 12b has an atmosphere rich in greenhouse gases, it could experience similar runaway heating, rendering it uninhabitable. Conversely, an atmosphere with too few greenhouse gases might fail to retain enough heat, making the planet too cold for liquid water.

Venus also demonstrates the impact of volcanic activity and atmospheric chemistry on a planet's climate. High levels of volcanic activity on Venus have likely contributed to its thick atmosphere and high surface temperatures. Understanding the geological activity on Gliese 12b could provide insights into its atmospheric composition and climate stability.

The comparison with Venus also underscores the importance of observational data in assessing habitability. The study of Gliese 12b's atmosphere will require detailed spectroscopic analysis to detect the presence and concentrations of greenhouse gases, water vapour, and other key components. This data will help determine whether the planet has a stable, Earth-like climate or if it is more akin to the hostile environment of Venus.

Chapter 6: Technological and Methodological Advances

6.1 Observational Techniques and Data Collection

The discovery of exoplanets like Gliese 12b relies on a variety of observational techniques, each offering unique insights into the properties of distant worlds. One of the most common methods is the transit method, which involves detecting the slight dimming of a star's brightness as a planet passes in front of it. This periodic dimming, or transit, provides valuable information about the planet's size, orbital period, and distance from its host star.

In the case of Gliese 12b, its discovery was facilitated by data collected from NASA's Transiting Exoplanet Survey Satellite (TESS). TESS scans large swathes of the sky, monitoring thousands of stars for signs of transiting exoplanets. By continuously observing these stars over extended periods, TESS can detect the subtle changes in brightness caused by orbiting planets, including those as small as Earth.

In addition to the transit method, other techniques have been employed to study Gliese 12b and its host star. Radial velocity measurements, for example, track the slight wobble of a star caused by the gravitational tug of an orbiting planet. By analyzing these wobbles, scientists can determine a planet's mass and orbital parameters.

Furthermore, direct imaging techniques have been used to study exoplanets by capturing their faint light directly. While challenging due to the overwhelming glare of the host star, direct imaging can provide valuable information about a planet's atmosphere, composition, and even surface features.

Contribution of TESS and Other Instruments

TESS has played a crucial role in advancing our understanding of exoplanets, including the discovery of Gliese 12b. Launched in 2018, TESS was designed to survey nearly the entire sky, focusing on nearby stars that are prime candidates for hosting exoplanets. Its wide-field cameras enable it to monitor hundreds of thousands of stars simultaneously, allowing for the detection of a wide range of planetary systems.

The high precision and sensitivity of TESS's instruments have led to the detection of numerous exoplanets, including small, rocky worlds like Gliese 12b. By continuously monitoring these stars, TESS provides valuable data on the frequency, size, and orbital characteristics of exoplanets across different stellar environments.

In addition to TESS, other ground-based and space-based observatories have contributed to the study of Gliese 12b and its star. The European Space Agency's (ESA) Gaia mission, for instance, provides precise measurements of stellar positions, distances, and motions, which are essential for characterising exoplanetary systems.

Furthermore, ground-based telescopes equipped with adaptive optics systems can correct for atmospheric distortions, allowing for sharper images of distant stars and their accompanying planets. These telescopes play a vital role in follow-up observations of exoplanets discovered by space-based missions like TESS.

6.2 Future Research and Technological Tools

The upcoming launch of the James Webb Space Telescope (JWST) represents a significant milestone in the study of exoplanets, including Gliese 12b. Scheduled for launch in 2021, the JWST is equipped with advanced instruments capable of analysing the atmospheres of distant worlds in unprecedented detail.

One of the key instruments onboard the JWST is the Near Infrared Spectrograph (NIRSpec), which will enable astronomers to study the chemical composition of exoplanet atmospheres by analysing their spectra. By observing how starlight filters through an exoplanet's atmosphere during a transit, NIRSpec can identify the presence of specific gases, such as water vapour, methane, and carbon dioxide.

Another instrument, the Mid-Infrared Instrument (MIRI), will complement NIRSpec by providing additional spectral coverage in the mid-infrared range. MIRI's capabilities are particularly well-suited for studying the thermal emission of exoplanets, which can reveal insights into their temperature, climate, and cloud cover.

Together, these instruments will revolutionise our understanding of exoplanet atmospheres, including the potential habitability of worlds like Gliese 12b. By detecting the signatures of key molecules and assessing atmospheric properties, the JWST will provide valuable constraints on the conditions present on distant exoplanets.

Spectroscopy and Atmospheric Analysis

Spectroscopy is a powerful tool for studying the atmospheres of exoplanets, offering insights into their chemical composition, temperature structure, and physical properties. By analysing the absorption and emission lines present in an exoplanet's spectrum, astronomers can infer the presence of specific gases and molecules.

For Gliese 12b, spectroscopic analysis will be crucial for determining the composition and characteristics of its atmosphere.

Observations with the JWST, in combination with ground-based telescopes, will allow astronomers to study the transmission and emission spectra of Gliese 12b as it transits its host star.

One of the key objectives of spectroscopic analysis is to identify potential biosignatures—chemical markers indicative of life—within exoplanet atmospheres. While no definitive biosignature has yet been identified, molecules such as oxygen, ozone, and methane are considered promising candidates. Detecting these molecules in the atmosphere of Gliese 12b could provide tantalising evidence for the presence of life beyond Earth.

Furthermore, spectroscopic observations can reveal insights into the climate and habitability of exoplanets by characterising their temperature profiles, cloud cover, and atmospheric dynamics. By comparing Gliese 12b's spectrum with theoretical models and observations of other exoplanets, astronomers can assess its potential for supporting life and identify areas for further study.

Chapter 7: Implications and Future Exploration

7.1 Astrobiological and Scientific Implications

The discovery of exoplanets like Gliese 12b has profound implications for astrobiology, the study of life in the universe. While the existence of liquid water is a crucial factor for habitability, it is not sufficient on its own to guarantee the presence of life. However, Gliese 12b's location within the habitable zone of its star and its rocky composition make it an intriguing candidate for further study.

One of the key questions in astrobiology is whether life exists beyond Earth and, if so, what forms it might take. The search for biosignatures—indicators of past or present life—plays a central role in this quest. On Earth, biosignatures include complex organic molecules, isotopic ratios, and the presence of specific chemical compounds associated with life processes.

For Gliese 12b, detecting biosignatures in its atmosphere could provide compelling evidence for the presence of life. Molecules such as oxygen, ozone, and methane are considered promising biosignatures, as they are produced by biological processes on

Earth. However, these molecules can also have abiotic origins, making their interpretation complex.

The challenge lies in distinguishing between biological and non-biological sources of these molecules. For example, oxygen can be produced through photosynthesis by plants and bacteria on Earth, but it can also be generated through abiotic processes such as photolysis of water or the breakdown of carbon dioxide. Similarly, methane can be produced by biological organisms, but it can also arise from geological processes like serpentinization.

To confidently identify biosignatures on Gliese 12b, astronomers will need to employ a combination of observational techniques and theoretical models. Spectroscopic analysis, particularly with instruments like the James Webb Space Telescope (JWST), will be crucial for detecting the presence of key molecules in the planet's atmosphere. However, interpreting these observations will require detailed knowledge of the planet's atmospheric chemistry, as well as consideration of potential abiotic sources.

Broader Impacts on Astrobiology

Beyond the search for life on individual exoplanets, discoveries like Gliese 12b have broader implications for our understanding of the prevalence and diversity of life in the universe. The sheer number of exoplanets detected to date—thousands and counting—suggests that planetary systems are common throughout the galaxy. This abundance of planets vastly expands the potential habitats for life beyond Earth.

Moreover, the diversity of exoplanetary systems challenges our preconceptions about what constitutes a habitable environment. While Earth-like planets within the habitable zone of Sun-like stars are of great interest, other types of planets, such as those orbiting red dwarfs like Gliese 12, could also harbour life under the right conditions. This diversity underscores the need for a comprehensive approach to astrobiology that considers a wide range of planetary environments and potential biosignatures.

The study of exoplanets also has implications for understanding the processes of planetary formation and evolution. By observing exoplanetary systems at various stages of development, astronomers can gain insights into the mechanisms that shape planetary architectures, from the initial formation of protoplanetary disks to the migration and evolution of planetary orbits.

Overall, the discovery of exoplanets like Gliese 12b opens new avenues for exploration and discovery in astrobiology. While the search for life beyond Earth remains challenging, each new exoplanet brings us closer to answering one of the most profound questions humanity has ever asked: are we alone in the universe?

7.2 The Future of Exoplanet Exploration

Despite the rapid progress in exoplanet discovery and characterization, significant challenges remain in studying distant worlds like Gliese 12b. One of the primary obstacles is the vast distances involved, which make direct observation and exploration challenging. Gliese 12b, for example, is located approximately 40 light-years away, making it inaccessible with current spacecraft technology.

Additionally, the faintness of exoplanets compared to their host stars presents challenges for observation. Even with advanced telescopes like the JWST, detecting and studying the atmospheres of exoplanets requires precise instrumentation and sophisticated data analysis techniques. Furthermore, the glare from the host star can overwhelm the faint signals from the planet, making it difficult to extract meaningful information.

Another challenge is the diversity of exoplanetary systems, which complicates efforts to identify common patterns and trends. While some exoplanets resemble those in our own solar system, many exhibit characteristics that are unlike anything seen before. Understanding the full range of planetary environments and their potential for habitability requires extensive observational data and theoretical modelling.

Long-term Goals and Prospects for Human Space Travel

Despite these challenges, the future of exoplanet exploration is bright, with ambitious goals and prospects for human space travel. Advancements in spacecraft technology, including propulsion systems and autonomous navigation, could eventually enable missions to distant exoplanetary systems.

One concept for reaching exoplanets is the use of solar sails, which harness the pressure of sunlight to propel spacecraft through space. Solar sail technology offers the potential for fast, efficient travel to nearby stars, albeit over long timescales. While current solar sail prototypes are experimental, ongoing research and development could lead to practical applications for interstellar travel in the future.

Another approach to reaching exoplanets is through the development of advanced propulsion systems, such as nuclear thermal propulsion or antimatter engines. These technologies offer the potential for much faster travel speeds than conventional chemical rockets, potentially reducing travel times to distant stars from centuries to decades.

In addition to robotic exploration, there is growing interest in the possibility of human missions to exoplanets. While such missions remain speculative at present, they represent a long-term goal for space exploration. The challenges of interstellar travel, including the effects of prolonged exposure to cosmic radiation and the logistical complexities of sustaining human crews over long durations, will require innovative solutions and international collaboration.

Ultimately, the study of exoplanets offers humanity a glimpse into the vast diversity of worlds beyond our own solar system. Whether through robotic exploration or eventual human missions, the quest to understand these distant planets and their potential for life represents one of the greatest scientific endeavours of our time. As technology continues to advance and our understanding of the cosmos deepens, the exploration of exoplanets will undoubtedly remain at the forefront of human exploration and discovery.